YOUR KNOWLEDGE HAS VALUE

- We will publish your bachelor's and master's thesis, essays and papers

- Your own eBook and book - sold worldwide in all relevant shops

- Earn money with each sale

Upload your text at www.GRIN.com and publish for free

Single-Phase Smart Protection Unit for Domestic Loads. Electrical Fault Protection

Silas Asiedu Asamoah

Kodzo Etornam Badza

Benjamin Amedormeh

GRIN

Bibliographic information published by the German National Library:

The German National Library lists this publication in the National Bibliography; detailed bibliographic data are available on the Internet at http://dnb.dnb.de.

ISBN: 9783346792952
This book is also available as an ebook.

© GRIN Publishing GmbH
Nymphenburger Straße 86
80636 München

All rights reserved

Print and binding: Books on Demand GmbH, Norderstedt, Germany
Printed on acid-free paper from responsible sources.

The present work has been carefully prepared. Nevertheless, authors and publishers do not incur liability for the correctness of information, notes, links and advice as well as any printing errors.

GRIN web shop: https://www.grin.com/document/1300294

UNIVERSITY OF ENERGY AND NATURAL RESOURCES

SCHOOL OF ENGINEERING
DEPARTMENT OF COMPUTER AND ELECTRICAL ENGINEERING

SINGLE-PHASE SMART PROTECTION UNIT FOR
DOMESTIC LOADS

By

ASAMOAH ASIEDU SILAS

BADZA ETORNAM KODZO

AMEDORMEH BENJAMIN

A Thesis Presented to the

University of Energy and Natural Resources

in Partial Fulfilment of the Requirements for the Degree of

Bachelor of Science

In

Electrical / Electronic Engineering

2019 University of Energy and Natural Resources

All rights reserved

DEDICATION

We would like to dedicate this project principally to the Almighty God for being our inspiration and strength throughout our stay in this university. Secondly, we dedicate this work to Mr. Forson Peprah of NEDco. We are very grateful for your immense contribution to the realization our plans. Finally, we recognise our families for their love and support throughout our years in school.

ACKNOWLEDGEMENTS

We would like to acknowledge the Almighty God who has been Our Guide and Inspiration throughout this project. Secondly, we express our sincere gratitude to our supervisor, Mr. Isaac Otchere for dedicating his time to ensuring that we choose and finetune our work to achieve maximum results. We also would like to acknowledge the Head of Department of Computer and Electrical Engineering, Dr. Eric Ofosu Antwi as well as the lecturers, who have imparted us with specialized knowledge over the years. Finally, we offer our heartfelt appreciation to Mr. Forson Peprah for his unflinching contribution, support and technical advice during this project right from the planning stages to the execution. We are eternally grateful.

TABLE OF CONTENTS

LIST OF FIGURES

LIST OF ABBREVIATIONS

GSM	Global System for Mobile Communication
ROCOFF	Rate of Change of Frequency
CML	Customer Minutes Lost
SCL	Serial Clock Line
SDA	Serial Data Line
SMS	Short Message Service
NEDCo	Northern Energy Distribution Company
ACS712	Alternating Current Sensor 715
HIL	Hardware-In-the-Loop
CT	Current Transformer
EC	Electric Circuit
IDE	Integrated Development Environment
LV	Low Voltage

ABSTRACT

In Domestic Electrical Power Systems, protection schemes are essential for the protection of equipment and lives, as well as for providing a reliable and continuous supply of energy. In this thesis, a single-phase fault protection unit for domestic loads is proposed. This proposed scheme uses intelligent components, mainly: a microcontroller, Global System for Mobile Communication (GSM), relay, current transformer, and circuit breaker to implement a smart electrical fault protection system.

The design is simulated and analyzed with the Proteus Design Suit Software and a prototype is designed and tested. The successful implementation of the smart protection system prevents damage to domestic appliances.

Keywords: Single-Phase fault detection, Fault Protection Unit, Smart Protection Unit

CHAPTER I

INTRODUCTION

1.0. Background

An electrical power system is an assembly of electrical components deployed to supply, transfer and use electric power. An electrical load also referred to as domestic load in a household setting, is any electrical device or portion of a circuit that consumes electric power (Wikipedia). However, the domestic load is collectively referred to as the total energy consumed by electrical appliances in household work. An electrical protective unit is a combination of electrical devices employed to protect equipment and other devices against abnormal, unreliable, and unpredictable conditions that may alter system values or cause damage to domestic equipment. Protection of domestic loads has become necessary because of the occurrence of short circuits, overloading of equipment, abnormal variations in voltage supply, etc. Failure to protect domestic loads may result in severe electric shock, fire outbreaks, loss of lives, and further damage to property. Various protection schemes have been designed to prevent and reduce these costly effects. This study introduces a new and smart way of protecting domestic loads.

1.1. Problem Statements

Domestic electrical faults cause severe damage to life and property. Over the years, numerous incidences of domestic fires have been reported across the length and breadth of this country most of which has been attributed to electrical faults. Several protective schemes have been developed over the years most of which not so budget friendly for the average Ghanaian household.

1.2. Objectives of Study

The objectives of this project are:

- To design and implement a smart single-phase protection unit that is capable of reporting faulty domestic conditions remotely via SMS.
- To implement an algorithm for smart fault detection and control system using Arduino IDE and simulate in Proteus IDE.
- To develop a prototype and test a smart fault detection and control system.

1.3. Scope of work

This project is designed to detect and report short circuit faults when they occur in a single-phase configuration for domestic loads. It makes use of a microcontroller, a contactor (in place of a circuit breaker), a GSM, a relay, current transformers, an LCD, and a load.

1.4. Thesis Organization

This section describes the main elements of the thesis. This study consists of five chapters and are elaborated on below:

- Chapter I defines key terms related to this work and introduces protection for domestic loads.
- Chapter II reviews various schemes developed for domestic load protection.
- Chapter III contains the system design and construction procedure.
- Chapter IV analyses the results and performance of the system.
- Chapter V contains the conclusion and recommendations for future work.

CHAPTER II

LITERATURE REVIEW

2.0. Details of Relevant Theory

Protection in electrical systems is defined as the science, skill, and art of applying and setting relays and/or fuses to provide maximum sensitivity to faults and undesirable conditions but not to avoid their operation on all permissible or tolerable conditions [1]. It is a very important aspect of electrical installations and must be done properly. The protection scheme has the function of safeguarding the system, minimizing damage and repair costs when faults occur, and ensuring the safety of users. It must also possess the following qualities:

- Selectivity: the ability to detect and isolate the faulty part only.

- Sensitivity: the ability to detect even the smallest fault and operate at its right setting before any irreparable damage is done.

- Speed: the ability to operate in a timely manner.

- Stability: the ability to ensure continuity of all healthy parts.

- Dependability: the ability to trip when the need arises.

- Security: the ability to not trip when not necessary.

The protective devices normally used in electrical and electronic devices are Fuses, Miniature Circuit Breakers (MCBs), Earth Leakage Circuit breakers (ELCBs), and Earthing or Grounding.

2.1. Review of Past Work on Fault Detection Methods

2.1.1. Household *Faults*

2.1.1.1. Voltage *Sags and Swells*

Voltage sag or dip is a short-duration reduction in RMS which can be caused by a short circuit, overload, or starting of electric motors [2]. Like electrical surges, sags and dips in electrical supply can often be attributed to devices connected to your power grid that are faulty or made with substandard materials and draw a lot of power when they are turned on.

Voltage swells are the opposite of dips and describe surges in voltage of 10% or more above normal or recommended usage. They can cause problems with appliances and overall power quality in a system. Swells can occur when a large load is turned OFF and the voltage on the power line increases for a short period of time. These affect the overall performance of various household loads such as air conditioning compressors, lighting loads, sensitive equipment like computers, etc. [3].

There are several factors that cause voltage sags or swells most of which occur at the secondary distribution line although the effect is felt by the customer. When a line-to-ground fault occurs, there will be a voltage sag until the protective switch gear operates [2]. Sudden load changes or excessive loads can cause a voltage sag [4]. Voltage sags can arrive from the utility grid, but most are caused by in-building equipment. In residential homes, voltage sags are sometimes seen when refrigerators, air-conditioners, or furnace fans start up.

2.1.1.2. Earth *faults*

Earth Fault is an inadvertent fault between the live conductor and the earth. When an earth fault occurs, the electrical system gets short-circuited and the short-circuit current flows through the system. The fault current returns through the earth or any electrical equipment, which damages the equipment. Such faults can cause objectionable circulating currents or may energize the housings of equipment at a dangerous voltage. It also interrupts the continuity of the supply and may cause electrical shocks to the user. To protect the equipment and for the safety of people, fault protection devices are used in the installation. Normally earth fault relays, earth leakage circuit breakers, ground fault circuit interrupters, etc. are used to restrict the fault current [5].

2.1.1.3 Arc Faults

An arcing fault is a dangerous phenomenon that is produced in the residential electrical system [6][7]. In many cases, the arcing phenomenon can produce losses, damages in the system or wire, injuries, death, or cause a fire [6]. Many approaches for detecting arcing faults have been considered for domestic applications in the following papers [8][9][10][11]. Arc faults are more difficult to detect due to domestic appliances that have a current signature [9] like the signature of an arcing fault [7][10] and in the case of the masking loads on electric lines [11].

2.1.1.4. Short Circuit Faults, Open Circuit Faults, and Overload

An open circuit fault is a type of fault that occurs because of a break in a circuit. It is a type of circuit in which there is no current flow which can occur because of wire breakage [12].

Overload faults are a result of a device doing work more than its rated capacity. When this happens, the device/machine will be drawing current more than its rated capacity which will result in a higher temperature in the device or machine and over time might cause it to overheat leading to a fire outbreak if there is no breaker in the circuit.

A short circuit is an abnormal connection between two nodes of an electric circuit intended to be at different voltages. This results in an electric current limited only by the Thevenin equivalent resistance of the rest of the network which can cause circuit damage, overheating, fire, or explosion [13]. In a short-circuited system the resistance of the circuit reduces drastically giving way for excessive current to flow in the circuit.

A short circuit is usually disastrous because of the amount of current flowing which might damage other components parts of the system and causes the temperature of the circuit to increase which can also lead to the formation of an electrical arc flash. Several protective and preventive measures have been considered and proposed over the years by several others. Amongst the several reviewed, (S. Rathor et al) presents the behaviour of a system under fault condition and directs us to the design of a circuit breaker [14]. (A. Apostolov et al) also looks at ways of Reducing the Effects of Short Circuit Faults on Sensitive Loads [15].

2.1.2. Protection Equipment

2.1.2.1 Circuit Breaker Using HIL Strategy

A Circuit breaker is an automatically operated electrical switch designed to protect an electrical circuit from damage caused by excess current from an overload or short circuit. Its basic function is to interrupt current flow after a fault is detected [16].

Typically, a breaker contains an electromagnet through which the current flows. If that flow exceeds a set level, the electromagnet becomes sufficiently energized to throw a mechanical switch, which breaks the circuit. A circuit breaker responds faster than a fuse and can also be reset manually instead of having to be replaced [17].

The various types of circuit breakers, their advantages and specific applications are discussed in [18][19][20]. (C Franck et al) also considers an overview of (High-Voltage Direct Current Circuit breakers) HVDC CBs, to identify areas where research and development are needed [21]. According to (J. Lezama et al), this method concerns the modelling of a typical electrical home installation which includes an electrical power supply, typical loads and series and parallel arc faults. This method of household protection considered two different loads (vacuum cleaner and kettle), and an arc in series on the electrical line [22]. The comparison between these models and real electric signals was done using harmonic analysis.

Finally, an arc-fault circuit breaker was proposed using Hardware-In-the-Loop (HIL) simulation approach.

The figure below from (Lezama et al., 2012) represents the domestic electrical network composed of a power supply, electrical wires, circuit breaker arc fault circuit interrupter and different loads (vacuum cleaner, drill, PC, kettle, fluorescent lamp…etc).

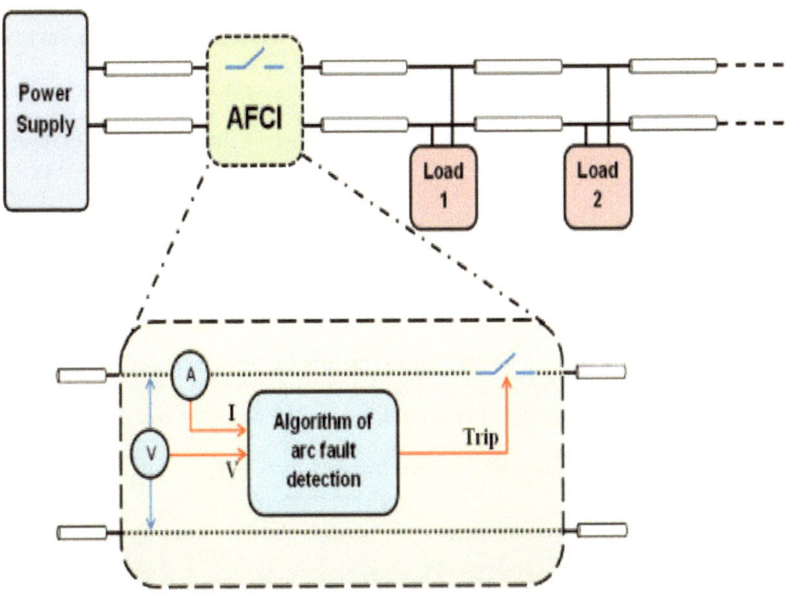

Figure II-1: Electrical line and circuit breaker for arc fault detection.

Image Source:

Lezama, J., et al. "Modeling of a Domestic Electrical Installation to ARC Fault Detection." *2012 IEEE 58th Holm Conference on Electrical Contacts (Holm)*, 2019, https://doi.org/10.1109/holm.2012.6336592.

This method has the advantage of ensuring the real functioning of a numerical detection algorithm by exploiting the models of the electric system to be protected. A major challenge with this method is its inability to detect other faults aside high arcs that may result in fire outages [22].

2.1.2.2 Overcurrent Protection with Electric Fuse

An electric fuse is an electrical safety device that operates to provide overcurrent protection of an electrical circuit (Wikipedia). It is made of a metal wire of strip that melts when too much current flows through it, thereby interrupting the current [23]. All fuses' posses a reverse characteristic of pre-arcing time or current that ranges from a definite minimum current level, below which they achieve equilibrium and can operate. The characteristic curve of these fuses because of this property makes it almost impossible to adjust the curve to improve the overall performance of the fuse. This makes it difficult in some applications to obtain the proper discrimination between protection electrical apparatus within the whole range of fault currents [24].

This special method of overcurrent protection according to (A. Plesca et al), implements a new type of fuse based on the concept of controlled fusing. The necessary energy to obtain the controlled fusing effect is provided by a current transform (CT) in connection with an adaptation electric circuit.

As represented in the figure below, the switch SW1 through the main circuit breaker (CB), supplies the autotransformer (ATR) which provides an adjustable voltage on the primary side of the current source (CS). A high adjustable current can be obtained on the secondary side of this electromagnetic device to perform experimental tests for both classical fuses and new fuse (FCF). The power for the electrode of the controlled fuse, is obtained from the secondary of the auxiliary transformer (AT) through the electronic switch (SW2).

The controlled fuse has an advantage of a very low cut-off current values compared to classical

fuses although this method of fuse control is suited to blade-contact-type fuses [24].

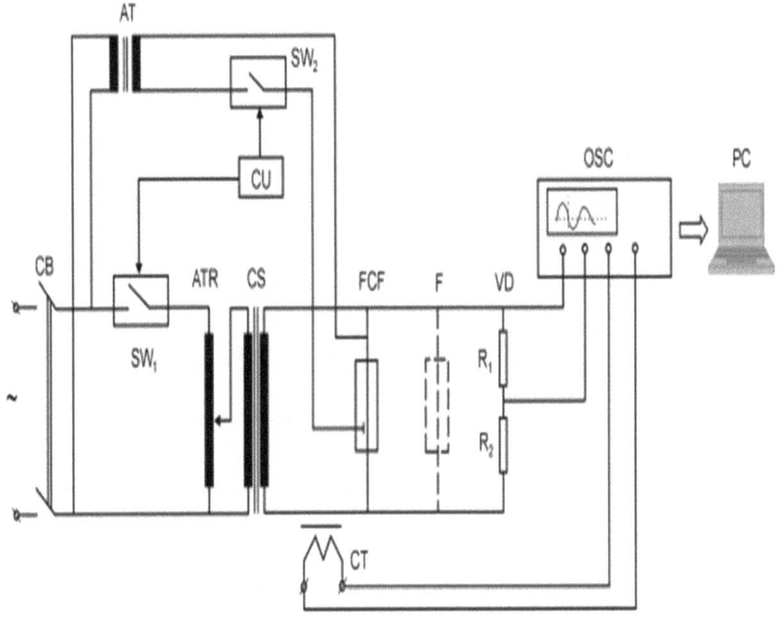

Figure II-2: Experimental Basic Diagram of a Controlled Fuse.

Image Source:

Plesca, Adrian, et al. "Overcurrent Protection Using a New Type of Electric Fuse." *2016 International Conference and Exposition on Electrical and Power Engineering (EPE)*, 2016, https://doi.org/10.1109/icepe.2016.7781321.

2.1.2.3 Smart Fault Detection Using WSN

Wireless Sensor Networks (WSNs) can be defined as a self-configured and infrastructure-less wireless networks to monitor physical or environmental conditions [25]. The wireless available technologies are Bluetooth, Wi-Fi, Wi-Max, wireless HART, Bluetooth and Zigbee [26]. Wireless sensor networks (WSN) are mainly designed for specific applications. several papers exist on the inculcation of these communication networks in electrical fault reporting.

According to (B. Uddin et al) electrical faults are detected and reported by the implementation of wireless monitoring and detection, using ACS712 Hall effect current sensors because of its better compatibility and quick response time. It makes use of the Zigbee module configured by XCTU software. The current sensors sense the current values of different loads, then the Wireless Sensor Network establishes a network between the monitoring room and the process to be monitored. This method however works for a single-phase system only [27].

According to [J.R Rana et al], voltage in the overhead line is continuously sensed using phase voltage sense sequence. When fault occurs, the voltage and current values deviate from their nominal ranges, the microcontroller analyses the type of fault then a relay connected to it operates. The GSM allows an SMS text to be sent to the one responsible in the control or monitoring room and the fault is immediately cleared. The fault clearing is done by protective devices like the relays and circuit breakers [28]. Unlike ZigBee, GSM has the advantage of a relatively wider network coverage, although a SIM card is needed for communication which comes with extra cost of communication.

CHAPTER III

SYSTEM DESIGN

3.0. Circuit Description

The block diagram of the proposed smart fault detection system is shown in Figure III-1. The

schematic diagram is made up of a current transformer, contactor (circuit breaker), relay, LCD,

Arduino micro controller, GSM shield, 9v battery and LV distribution lines serving load.

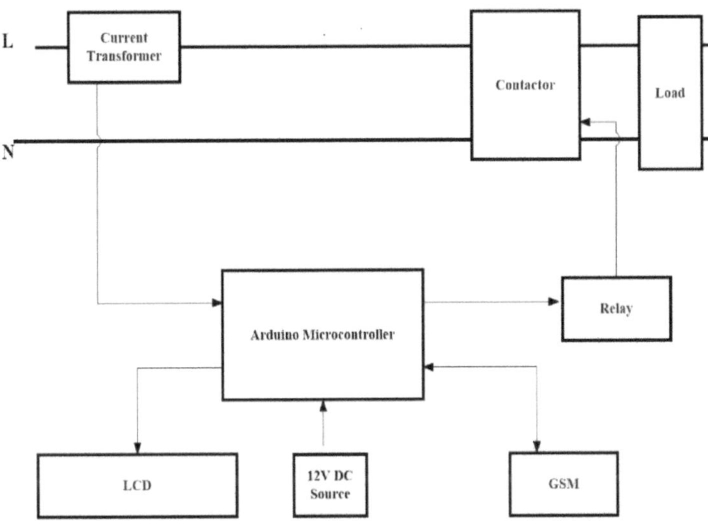

Figure III-1:Block Diagram of Fault Detection System.

[Author's Own Work]

The circuit diagram shows the connection of various components of the fault detection system.

The GSM is mounted on the Arduino microcontroller and has pins 7, 3 and 2 on the Arduino

UNO reserved for its use, with the relay connected to pin 6. The LCD receives its supply from

the 5V port of the Arduino and has its ground connected to the Arduino ground. The SCL and

SDA ports of the LCD connect to their respective ports on the Arduino. The current transformer is clamped around the 240V supply distribution line and has its secondary together with its circuitry (made up of a 28-ohm burden resistor, two 100k ohm resistors and a 100μF, 10V capacitor) connected to the A0 port of the Arduino. There is a contactor connected to the relay and across the distribution lines, with varying loads interchangeably connected at its end to simulate the fault condition.

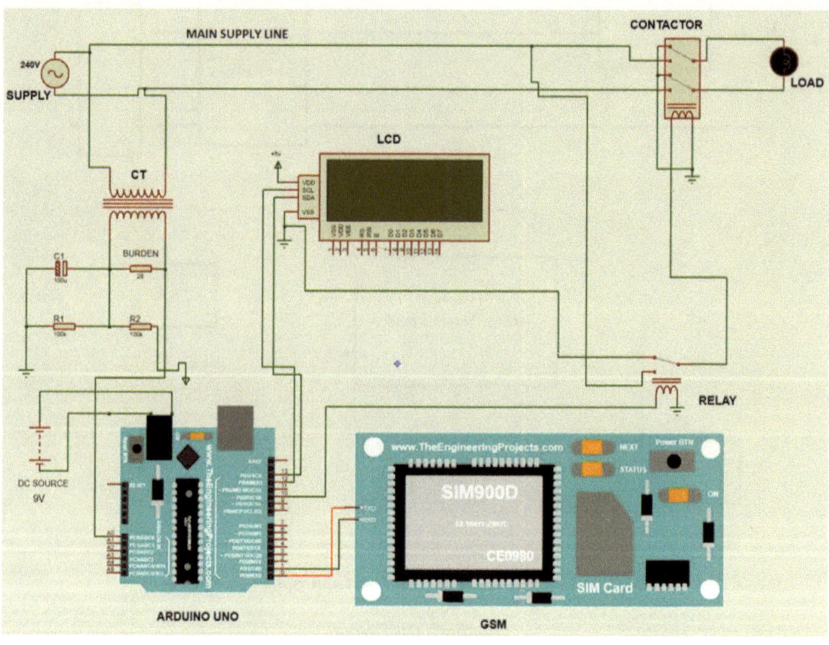

Figure III-2:Circuit Diagram of Fault Detection System with Proteus 6

[Author's Own Work]

3.1. Principle of Operation

The smart fault detection and control system is divided into two main parts, i.e., the control circuitry and the power circuitry. The control circuitry involves control, coordination, and reporting systems whereas the power circuitry carries the load current and isolates the system during a fault condition.

In the event of fault or overload, the line current increases beyond the design load current and causes the contactor to trip ensuring safety to the household appliances. If the fault current is allowed to flow over some time, the possibility of damaging connected appliances or causing a fire outbreak is very high. Under normal load condition no report is generated but under faulty condition the abnormal current magnitude causes the breaker to trip and a report is generated automatically indicating the fault condition. This report is sent to the customer in the form of an SMS indicating the fault magnitude and time of occurrence. The current in the line is continuously measured by the current transformer. The Arduino microcontroller is responsible for coordination and control of the various subsystems like the relay control and report generation and transmission.

The Circuit breaker is responsible for opening and closing of the load circuit under the command of the Arduino microcontroller through a relay.

The LCD displays the state of the system visually at any point in time for easy on-site monitoring. The operator has the ability of turning on or off the load remotely even in the absence of a fault via SMS.

3.2.   Selection of Components

3.2.1 Circuit Breaker (Contactor) and Overload Relay

A circuit breaker is an electromechanical switch that operates under both normal and abnormal
conditions. That means they are made to carry both normal and faulty load currents. For this
work, a contactor that works almost the same as a circuit breaker was used due to unavailability
of a miniature circuit breaker that works under magnetism. Contactors however do not have
arc quenching medium and that limits their application in larger current circuits.

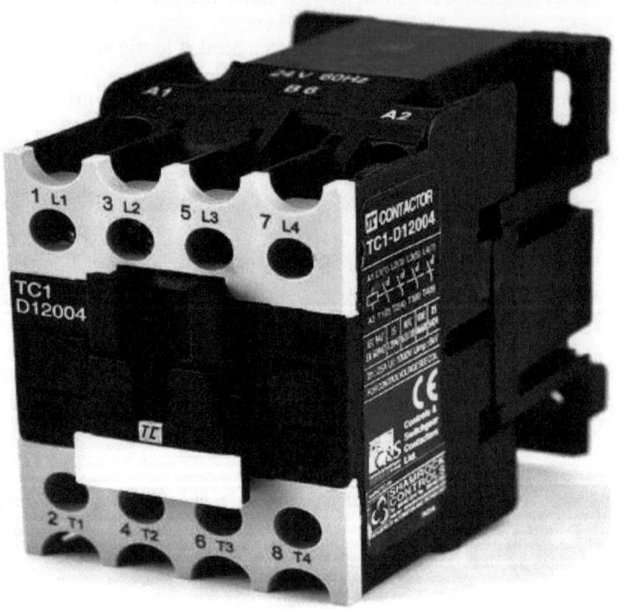

Figure III-3: A Contactor in place of a Circuit Breaker.

Image Source:

"3 PC Board C&S Electrical Power Contactor." *Indiamart.com*, 2019,
https://www.indiamart.com/proddetail/c-s-electrical-power-contactor-16492272412.html.

3.2.2 Relay Circuitry

Relays are typically used to switch a device whose voltage and current cannot be handled directly by microcontrollers. A relay is an electromagnetic switch which mechanically closes an electric circuit.

A new relay has emerged; this is a solid-state relay which does not follow the ideal operational principles as the electromagnetic relay [29]. Most heat sources and electrical motors are controlled directly with relays. Theses relays find applications in Domestic appliance, office machine, audio, equipment, automobile, etc [30].

Figure III-4: Two Channel Optocoupler relay.

Image Source:

"2 Channel 5V Relay Module with Optocoupler (DC 30V/10A, AC 250V/10A): Sharvielectronics: Best Online Electronic Products Bangalore." *Sharvielectronics*, 22 Aug. 2022, https://sharvielectronics.com/product/2-channel-5v-relay-module-with-optocoupler/.

3.2.3 Current Transformer

The current transformer with ratio 100A:50mA measures the current flowing through a conductor and reproduces in its secondary winding a current that is proportional to that in its primary winding. It provides safety for the relay by producing lower-level relay inputs.

To be able to use this current transformer, an accompanying circuit comprising of resistors and a capacitor is built. The figure below shows the current transformer used.

Figure III-5: Current Transformer.

Image Source:

"AC Current CLAMP YHDC SCT-013-000 100a/50ma Split Core CT." *Ac Current Clamp Yhdc Sct-013-000 100a/50ma Split Core Ct - Buy Ac Current Clamp Sct013,Split Core Ct Sct013,Sct013 Product on Alibaba.com,* Dechang Electronic Co.Ltd , 2019, https://yhdc.en.alibaba.com/product/60510732963-218696501/AC_current_clamp_YHDC_SCT_013_000_100A_50mA_split_core_CT.html.

3.2.4 Liquid Crystal Display

Liquid Crystal Display (LCD) depicted in figure 3.8. The 20x4 LCD unit is a very basic module and it is commonly found in many devices and circuits. The module has a lot of advantages over seven segments and other multi segment LEDs. It is economical, easily programmed, has no limitation of displaying special and even custom characters, animations and so on. It can display 20 characters per line and there are 4 such lines. This is the reason why it is called 20x4. [31].

Figure III-6:20x4 Liquid Crystal Display.

Image Source:

Old_Grey. "Part Request Blue Backlight 20X4 Display." *Fritzing Forum*, Fritzing , 13 Apr. 2019, https://forum.fritzing.org/t/part-request-blue-backlight-20x4-display/4617/6.

3.2.5 Communication Equipment (Arduino GSM Shield V2)

A very key function of this system is its ability to send and receive SMS control commands wirelessly. This is possible in the Arduino environment using GSM Shield to create data connectivity to the broad spectrum of the world today (internet and the voice/data using cellular network). The Arduino GSM Shield 2 is the heart of this project, it enables the Arduino board to be connected to the user, by providing SMS communication. The shield has quad-band GSM/GPRS modem that works at frequencies GSM850MHz, GSM900MHz, DCS1800MHz and PCS1900MHz. It also supports TCP/UDP and HTTP protocols through a GPRS connection and has speed (downlink and uplink) of 85.6 kbps. A SIM is required to interface the shield and with the cellular network [32]. The Arduino GSM shield V2 is shown in the Figure 3.9 below.

Figure III-7: GSM Arduino Shield

Image Source:

"Arduino GSM Shield 2 (Antenna Connector)." *Tech Bazar || টেক বাজার,* 9 Sept. 2019, https://www.techbazar.com.bd/product/arduino-gsm-shield-2-antenna-connector/.

3.2.6 Arduino UNO

The Arduino UNO is an open-source microcontroller board based on the Microchip ATmega328P microcontroller and developed by Arduino.cc [33]. It has digital and analogue input/output (I/O) pins that may be interfaced to various expansion boards (shields) and other circuits. There are 14 Digital pins, 6 Analog pins which are programmed with the Arduino IDE (Integrated Development Environment) via a type B USB cable. The Arduino board is usually powered by a USB cable or by an external 9-volt battery and it accepts voltages between 7 and 20 volts. In our case, it is powered by an external battery. It is used to control actions and responses of various components in the system, as programmed.

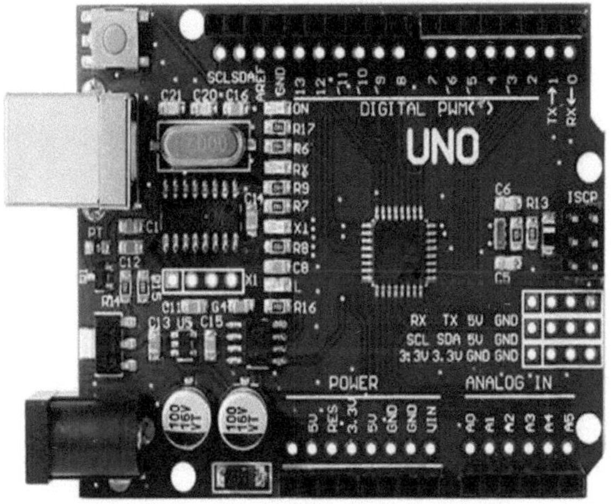

Figure III-8:Arduino-UNO-R3

Image Source:

Sattar, Hina, et al. "An Intelligent and Smart Environment Monitoring System for Healthcare." *Applied Sciences*, vol. 9, no. 19, 2019, p. 4172., https://doi.org/10.3390/app9194172.

3.2.7 Load

Loads will be interchangeably connected to simulate for short circuit faults. For the purposes of this work, any load that draws a current of more than 5A causes the contactor to open, isolating it from the rest of the system. For demonstration purposes, a lamp serves as the normal load and an iron serves as the fault current-causing load. [34]

3.3. System control algorithm and Simulation

Two main software are used for simulating and programming the microcontroller for this project. Arduino IDE 1.8.2 provides the interface to write to the microcontroller to program it. The Arduino IDE uses the C++ programming language.

The system simulation was done in proteus IDE 8. Proteus combines its internal components and Arduino code for effective simulation. The smart fault detection and control algorithm used in this thesis is depicted in the appendix.

CHAPTER IV

RESULT AND DISCUSSION

4.1. Simulation, Construction and Testing

The project was successfully implemented after successful simulation. The Arduino micro controller code (algorithm) was written with Arduino IDE 1.8.2. The simulation was done in proteus IDE (proteus 8). The components that were considered in the simulation are as follows: Distribution transformer which supplies the electrical power to various load was replaced by a generator for the purposes of the simulation. This was necessary for simplicity; current transformer which continuously measures the line current at the LV feeder of the transformer, distribution line that interconnects the transformer (generator in this case) and the load electrically, Arduino micro controller that coordinates and controls the entire process, Arduino GSM shield is responsible for communication to the operator; this device was made to communicate with a mobile phone by sending ripple control and reports, a 16 x 2 LCD display responsible of displaying all the actions of the processes, relay used in this work was replaced by a contractor for the purpose of handling higher current than normal for relay, control wires were used for connecting the individual components with the exception of the power circuit. Figure 4.11 depicts the simulation results. Respective line current (R, Y, B) were displayed on the LCD. For protection control, the line current was limited to 5A, this means that any current higher than this threshold was considered as overload or short circuit, and this causes the contactor to open to save the system thereby protecting the transformer and the line.

The simulation results confirmed the physical construction of the prototype. The entire hardware (physical) of the work was housed in a junction box. The figures below show the system simulation with Proteus and the final working system construction.

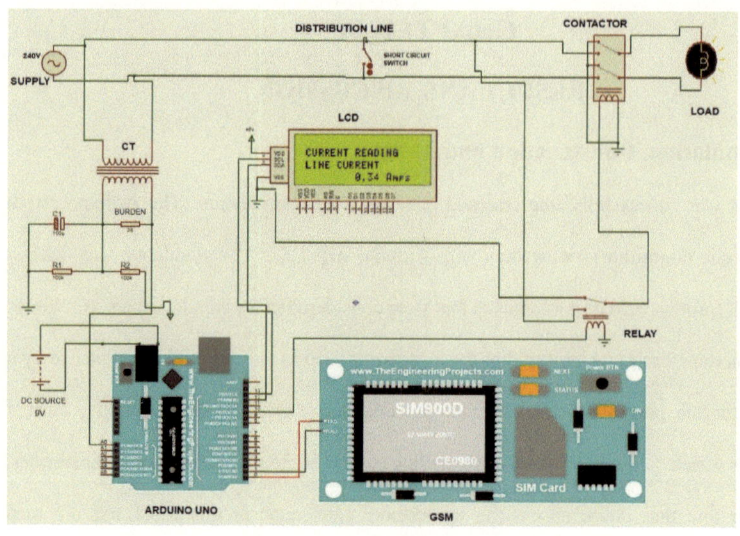

Figure IV-1: System Simulation with Proteus Showing Normal Current Reading.

[Author's Own Work]

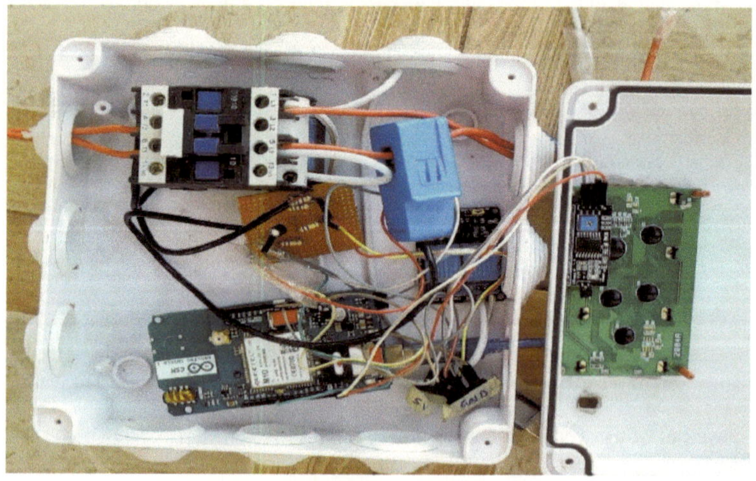

Figure IV-2: Final Construction Showing Internal Components.

[Author's Own Work]

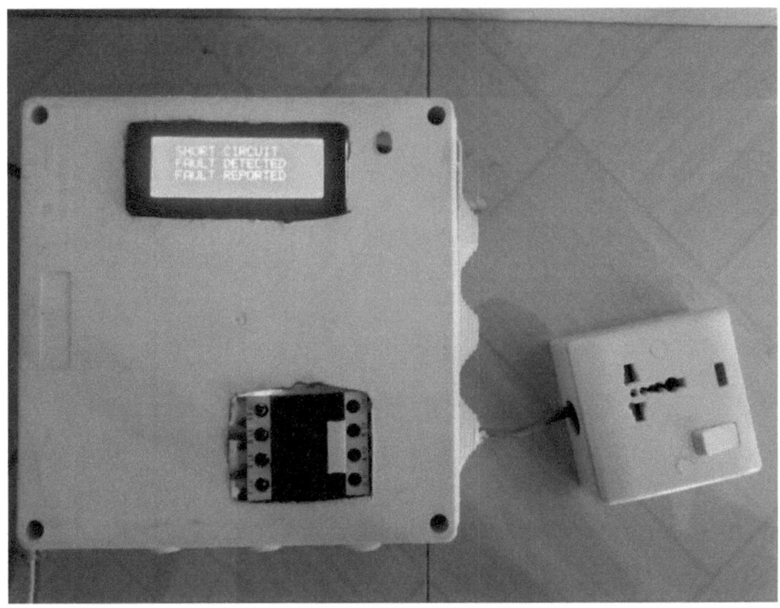

Figure IV-3: Working Final Construction Showing Fault Condition.

[Author's own Work]

The system above was constructed according to the methodology highlighted earlier in the previous chapter. It consists of a junction box which houses all the individual components. There is an Arduino microcontroller, GSM module, contactor, relay, current transformer, a socket for connection of a load.

When the system was switched on and under normal conditions, the LCD displayed the normal conditions. However, when a load with a current rating higher than the programmed default current rating was connected to the system, that load was automatically switched out of the system. An SMS was sent to the operator indicating the faulty condition. When the fault was cleared, the operator reconnected the system with an SMS, made possible by the GSM.

Whenever the load, an electric iron in this case (to mimic the fault current), was connected, the system tripped again isolating the iron.

CHAPTER V

CONCLUSION AND RECOMMENDATION

5.0 Summary of Main Study

The system as designed and tested, detects short circuit faults, disconnects the system from the fault and sends an SMS with information about the fault to the operator. The operator is also able to reconnect the system via SMS text, when the fault is cleared.

5.1 Directions for Future Research

Future work may consider using rechargeable batteries to power the microcontroller. These batteries could be charged by the main line during healthy conditions via a rectifying circuit. Open circuit fault detection could also be considered.

REFERENCES

[1]   T. J. D. J. Lewis Blackburn, "Introduction and General Philosophies," in Protective Relaying: Principles and Applications, Fourth Edition, 4th ed., Taylor and Franscis Group, Ed. London New York: CRC Press, p. 4.

[2]   M. H. . Bollen, "solving power quality problems," IEE Press, no. Voltage Sags and interruptions, p. 139, 1999.

[3]   S. Saksena, B. Shi, and G. Karady, "Effects of voltage sags on household loads," pp. 1593–1598, 2005.

[4]   M. H. Kazibwe, Wilson E., Sendaula, "Electric power quality control techniques," V. N. Reinhold, Ed. New York, 1993, p. 11.

[5]   T. Point, "Electrical safety earth fault protection," Tutorials point, 2019. .

[6]   "Residential building electrical Fires. United States Fire Administration," FEMA 2008, Topycal Fire Rep. Ser., vol. 8, no. 2.

[7]   J. J. Shea, "Identifying causes for certain types of electrically initiated fires in residential circuits," Fire Mater., vol. 35, no. 1, pp. 19–42, 2011.

[8]   W. J. L. et Y. C. Li, "Arc fault detection based on wavelet packet," in Machine Learning and Cybernetics, 2005, pp. 1783–1788.

[9]   B. L. et Y. Rebière, "Détection d'arcs électriques séries par analyse temps-fréquence et traitement morphologique," 2001.

[10]   et Z. R. N. Hadziefendic, M. Kostic, "Detection of series arcing in low-voltage electrical installations," Eur. Trans. Electr. Power, vol. 19, no. 3, pp. 423–432, 2009.

[11]   C. E. Restrepo, "Arc fault detection and discrimination methods," in Electrical contacts-2007, the 53rd ieee holm conference, 2007, pp. 115–122.

[12]   O. Quadri, "Open Circuit, short circuit and overload fault," 2019. .

[13]   wikipedia contributors, "short circuit," wikipedia the free encyclopedia, 2019. .

[14]   S. Rathor, "Short Circuit Analysis Case Study & Circuit Breaker Design Master in Power System," no. April 2011, 2016.

[15]   A. Apostolov, "Reducing the Effects of Short Circuit Faults on Sensitive Loads in Distribution Systems," Technology, pp. 1–12.

[16]   wikipedia contributors, "Circuit Breaker," wikipedia the free encyclopedia, 2019. .

[17]   T. Economist, "Smart circuit breaker for energy efficient homes." .

[18]   Z. Ganhao, "Study on DC circuit breaker," Proc. - 2014 5th Int. Conf. Intell. Syst. Des. Eng. Appl. ISDEA 2014, pp. 942–945, 2014.

[19]   M. Knezev, Z. Djekic, and M. Kezunovic, "Automated circuit breaker monitoring," 2007 IEEE Power Eng. Soc. Gen. Meet. PES, no. April 2014, 2007.

[20]   L. Rubino, G. Rubino, P. Marino, L. P. Di Noia, and R. Rizzo, "Universal Circuit Breaker for PV power plants," 2017 6th Int. Conf. Clean Electr. Power Renew. Energy Resour. Impact, ICCEP 2017, pp. 750–755, 2017.

[21]   C. M. Franck, "HVDC circuit breakers: A review identifying future research needs," IEEE Trans. Power Deliv., vol. 26, no. 2, pp. 998–1007, 2011.

[22]   J. Lezama, P. Schweitzer, S. Weber, E. Tisserand, and P. Joyeux, "Modeling of a domestic electrical installation to arc fault detection," Electr. Contacts, Proc. Annu. Holm Conf. Electr. Contacts, 2012.

[23]   C. Wikipedia, "fuse," wikipedia, 2019. [Online]. Available: https://en.wikipedia.org/wiki/Fuse_(electrical). [Accessed: 18-Jun-2019].

[24]   A. Plesca, C. Dumitrescu, G. Zhang, and D. Han, "Overcurrent protection using a new type of electric fuse," Proc. 2016 Int. Conf. Expo. Electr. Power Eng. EPE 2016, no. Epe, pp. 143–146, 2016.

[25]   M. . Paridah, A. Moradbak, A. . Mohamed, F. abdulwahab taiwo Owolabi, M. Asniza, and S. H. . Abdul Khalid, "Overview of wireless sensor network," Intech, vol. i, no. tourism, p. 13, 2016.

[26]   A. Kulkarni and A. Patange, "A Review on Zigbee, Gsm and Wsn Based Home Security by Using Embedded Controlled Sensor Network," Int. J. Embed. Syst. Appl., vol. 6, no. 3/4, pp. 01–08, 2017.

[27]   B. Uddin, A. Imran, and M. A. Rahman, "Detection and locating the point of fault in distribution side of power system using WSN technology," 4th Int. Conf. Adv. Electr. Eng. ICAEE 2017, vol. 2018-Janua, pp. 570–574, 2018.

[28]   P. A. Gulbhile, J. R. Rana, and B. T. Deshmukh, "Overhead line fault detection using GSM technology," IEEE Int. Conf. Innov. Mech. Ind. Appl. ICIMIA 2017 - Proc., no. Icimia, pp. 46–49, 2017.

[29]   E. Lab, "4 channel arduino relay module," electronics lab, 2019. [Online]. Available: http://www.electronics-lab.com/project/4-channel-relay-board/. [Accessed: 18-Jun-2019].

[30]   C. Songle, "SonglePower Relay SRD-5VDC-SL-C." pp. 1–2, 2000.

[31]   "Milestones: Liquid Crystal Display," IEEE Global History Network, 2011. [Online]. Available: https://en.wikipedia.org/wiki/Liquid-crystal_display. [Accessed: 04-Jun-2019].

[32]   A. Co-operation, "Arduino GSM Shield 2 (Integrated Antenna)," 2019. [Online]. Available: https://www.arduino.cc/en/Main.ArduinoGSMShield. [Accessed: 22-May-2019].

[33]   A. Co-operation, "Arduino Uno," Arduino cc, 2017. [Online]. Available: https://www.arduino.cc/en/Main/ArduinoBoardUno. [Accessed: 19-May-2019].

[34]   M. A. R. Uddin Bahar, Imran Asif, "Detection and locating the point of fault in distribution side of power system using WSN technology," 2017 4th Int. Conf. Adv. Electr. Eng., 2017.

Code

```
#include <LiquidCrystal_I2C.h>              // I2C lcd library
#include "EmonLib.h"                        // CT library
#include <Wire.h>                           //LCD library
#include <GSM.h>                            // GSM library


EnergyMonitor emon1;                        // create an instance of the CT
LiquidCrystal_I2C lcd(0x27, 20, 4);


#define PINNUMBER ""          //optionally add pin number of sim card GSM
GSM gsmAccess;
GSM_SMS sms;


char senderNumber[20];                      //to hold No. sending to sim in GSM
char remoteNum[]="0551506218";              //No. GSM is sending to
char inChar;                                // variable to store the incoming message
int R1 = 6;                                 // relay pin
int R_indicator = 12;                       //led for relay state indication
float maxCtVal =5;                          // above is fault.
float ctVal=0.00;                           // sensed current values to Arduino
boolean notConnected =true;
String unit = " Amps";
String op1 = "+233543714377";              // SMS from only these sources  only
String op2 = "+233551506218";
String op3 = "+233555818980";
boolean faultCheck = true;
int i = 1;
void initPins();void initLCD();void initGSM();
void show(String row1,String row2="",String row3="",String row4="");
void receiveSMS();  void sendSMS();void otherwise();
```

```cpp
void setup() {
 Serial.begin(112500);
 initLCD();
 initPins();
 digitalWrite(R1,HIGH);
 initGSM();
 emon1.current(0,150);                    //CT analogue pin and calibration
 digitalWrite(R1,HIGH);
 ctVal =0.00;
}


void loop() {                  // NORMAL CONDITION
  float ctVal = emon1.calcIrms(1480);
  if(ctVal <= maxCtVal){
    Serial.println(ctVal);
    digitalWrite(R1,LOW);                 //initial relay state: on
    show("CURRENT READING","","LINE CURRENT: ");
    lcd.setCursor(9,3);
    lcd.print(ctVal + unit);
    receiveSMS();
    faultCheck = true;   }
              // FAULT CONDITION
   else{
    while(faultCheck){
    digitalWrite(R1,HIGH);               // turn off relay

    while (i <4){                        // this loop sends the SMS three times only
    //sendSMS("Line Fault Detected. Line State: DISCONNECTED");
    delay(2000);
    show(String(i) + " :SMS SENT");
    i = i+1;
    }
```

```cpp
    show("SHORT CIRCUIT","FAULT DETECTED","FAULT REPORTED");
     if(sms.available()>0){
      receiveSMS();
      i = 1;                          // reset i to 1 for the next running condition
      faultCheck = false;             // condition to break out of the loop
      }
     else if(sms.available()<0){
      faultCheck = true;    }         // else stay here if no SMS is available
   }
  }
}
          //*   FUNCTION DECLARATIONS
void initPins(){
 pinMode(A0,INPUT);                   // CT-1 source pin
 pinMode(R1,OUTPUT);
}
void initLCD(){                       // Initialize the LCD
  lcd.init();   lcd.init();
  lcd.backlight();
  show("INITIALIZING","","PLEASE WAIT");
}

void initGSM(){                       // Initialize the GSM
 while(notConnected){
  if(gsmAccess.begin(PINNUMBER)==GSM_READY)
   notConnected = false;
  else{
   lcd.home();
   lcd.print("GSM NOT CONNECTED");
   delay(500);
  }
 }
 show("","GSM INITIALIZED");
```

```cpp
  delay(800);
}
            // Create function for displaying on the LCD
void show(String row1="",String row2="",String row3="",String row4=""){
 lcd.clear();
 lcd.setCursor(2,0); lcd.print(row1);
 lcd.setCursor(2,1); lcd.print(row2);
 lcd.setCursor(2,2); lcd.print(row3);
 lcd.setCursor(2,3); lcd.print(row4);        }
void sendSMS(String txtMsg){        // function for sending SMS
  sms.beginSMS(remoteNum);
  sms.print(txtMsg.c_str());
  sms.endSMS();
  show("","FAULT REPORT","SENT");
  delay(1000);
}


void receiveSMS(){                   // function for receiving SMS
        if(sms.available()){
           sms.remoteNumber(senderNumber,20);
           //lcd.print(senderNumber);
           show("CTRL SIGNAL","RECEIVED.");
           delay(2000);
          String sender = String(senderNumber);
   if (sms.peek() == '#' || (sender != op1 &&  sender != op2 &&  sender != op3)) {
           show("SMS status", "Discarded");
           sms.flush();
           delay(1000);
           return;
   }


  String cmd = "";                   // declare variable command
    while((inChar = sms.read()) != '\0'){
```

```cpp
    cmd += String(inChar);           // read and update the string msg
  }
  show("","COMMAND: "+ cmd);
  delay(2000);


  String msg ="";                    // declare new string called message
  if (cmd =="CLOSE"){                // check the received cmd and act accordingly
    digitalWrite(R1,LOW);
    show("LINE","CONNECTED");
    delay(2000);
    msg="Line Successfully RECONNECTED.";
    sendSMS(msg);
    sms.flush();
  }


  else if (cmd == "OPEN"){
    digitalWrite(R1,HIGH);
    show("LINE","DISCONNECTED");
    delay(1000);
    msg="Line Successfully DISCONNECTED.";
    sendSMS(msg);
    sms.flush();
  }
  else{
  show("SMS STATUS:","UNKNOWN COMMAND!");
  delay(1000);
  show("SMS status", "Discarded");
  sms.flush();               //Discard the message
  delay(1000); }
 }
 sms.flush();                //delete the msg from memory
}
}
```

YOUR KNOWLEDGE HAS VALUE

- We will publish your bachelor's and
  master's thesis, essays and papers

- Your own eBook and book -
  sold worldwide in all relevant shops

- Earn money with each sale

Upload your text at www.GRIN.com
and publish for free